THE FUTURE IS VERIFIED

The Digital Transformation of Property as We Know It

Elizabeth Sullivan

The illustrative narratives in this book are composite scenarios drawn from industry patterns and are not accounts of specific individuals.

First Edition

Author's Note

The personal stories you'll encounter in these pages, including Jen and Braeden's digital closing, Lexi's mystery lien, Melissa Rossi's transformation of her title office, Kalysta's AR billboard empire, and the road trip investment in Chapter 3, are all illustrative narratives. They are composite scenarios drawn from real patterns, challenges, and outcomes documented across the blockchain real estate industry, rather than accounts of specific individuals.

The factual claims in this book are drawn from publicly available sources including company press releases, government announcements, industry association research, and published news coverage. Where precise figures could not be verified against primary sources, they have been replaced. Sections marked with * involve financial actions. Nothing in this book constitutes financial advice. Consult a qualified financial advisor before making any investment decisions with directional language that preserves the underlying point.

A note on process: in researching and drafting this book, I made use of AI tools to assist with synthesizing research, stress-testing arguments, and refining complex concepts into accessible language. Every idea, opinion, and conclusion is my own. I believe transparency about how we work is part of the integrity this book advocates for. In the same way blockchain makes processes visible, I want you to see mine.

Table of Contents

Introduction

A Digital Shift in Real Estate

Real estate has a trust problem. Not a paperwork problem, not a technology problem, not even an efficiency problem (though it has all of those too). The deeper issue is that every real estate transaction in the world today depends on a chain of humans agreeing to trust each other: the escrow officer trusts the wire, the title company trusts the county database, the buyer trusts the seller's disclosure, the bank trusts the appraisal. When that chain holds, a home changes hands. When it breaks through fraud, error, delay, or simple miscommunication, deals collapse, money is lost, and people who should be celebrating are left anxious and exhausted.

For centuries, we've responded to that trust problem the same way: by adding more intermediaries. More verification steps. More insurance products to cover the gaps the verification misses. More lawyers to manage the disputes the insurance doesn't cover. Each new layer was a patch on the one below it, and the patches stacked up. Today's real estate transaction involves, on average, a dozen separate parties operating on incompatible systems, and it still takes between 30 and 60 days to close. We've been piling fixes on top of workarounds for so long that most people in the industry have stopped asking whether the foundation was right in the first place.

"Blockchain asks that question. And its answer is worth paying attention to."

What blockchain offers real estate is not another layer. It's a relocation of trust. Instead of trust residing in institutions, it resides in mathematics. A smart contract doesn't trust the wire transfer. It verifies it. A blockchain title registry doesn't trust the county database. It replaces it with a ledger that is cryptographically impossible to alter without detection. This is not a speed improvement. It's not a cost-reduction initiative. It's a change in what the system is built on.

That argument, that blockchain relocates trust rather than simply digitizing the process, is what this book is about. The professionals who understand what's actually changing (not just the tools, but the underlying architecture of ownership) will be the ones who shape what real estate looks like for the next generation.

When I first began researching this book, I went around asking top-producing brokers what they knew about blockchain. Most of them looked at me like I'd asked about astrology. The most common answer: "I'm not sure, it feels intangible. I tend to stick with more traditional investments." Totally fair response, honestly, and it reminded me of what I've heard about the internet in the early 1990s. Not skepticism exactly, more like a failure of imagination. The people who figured out early what the internet was actually changing gained advantages that compounded for decades. The same thing is happening in real estate right now.

This book is structured around five transformations: how transactions are executed, how titles are recorded, how property is owned and invested in, how virtual space is developing its own real estate economy, and how the physical world is gaining a digital layer that can itself be owned and monetized. Each chapter moves from the familiar problem to the specific way blockchain changes it, grounded in real case studies, concrete examples, and the honest acknowledgment of where these technologies are still maturing.

You don't need to be a technologist to follow along. The trust that once lived in institutions is migrating to mathematics. The question for every real estate professional, investor, and owner reading this is the same: how do you position yourself before that migration completes?

Now, let's begin.

Chapter One

Blockchain Powered Transactions

How trust moved from institutions to mathematics

Ask any homebuyer, seller, or real estate professional about closing day and you'll hear a mix of excitement and exasperation. It's a finish line that too often feels like a maze. There are the endless stacks of paperwork requiring signatures and notarizations, the long email threads with attachments that go missing just when you need them most, and the parade of intermediaries: escrow officers, lenders, lawyers, appraisers, title companies, all operating on separate systems, each with their own timelines and requirements.

The result is a fragmented process that demands constant coordination just to stay afloat. The average home closing in the United States takes between 30 and 60 days, and that's if everything goes smoothly. The system isn't broken, but it is undeniably outdated. It was built for an era of filing cabinets and faxes, not for an age of instant communication and digital systems.

Understanding Smart Contracts

To understand what blockchain changes, you first need to understand smart contracts, because they appear in nearly every chapter of this book and they're the mechanism through which most of blockchain's real estate applications actually work.

A smart contract is a self-executing agreement stored on the blockchain. It doesn't need a notary, an escrow officer, or a paralegal to enforce it. Instead, it runs on logic: if a set of predefined conditions are met (payment received, inspection approved, title verified), the contract automatically completes the next step. The rules are enforced by code, precise, consistent, and it doesn't take lunch breaks.

In a blockchain-enabled real estate transaction, escrow funds are held not by a third-party account but by the smart contract itself, released automatically the moment both parties meet their obligations. Title checks and ownership verification occur in seconds because they reference data already secured on the blockchain. You'll encounter smart contracts throughout this book. Same idea every time: trust encoded in mathematics rather than handed off to a person. In 2015, Ethereum introduced smart contracts to the world. Real estate has been slowly catching up to what that actually means ever since.

From Paper Closings to Blockchain

For centuries, real estate has run on paper. The harder problem wasn't the technology. It was the law. A blockchain can transfer a token instantly. What it cannot do on its own is satisfy a county recorder's office, a title insurer, or a mortgage lender. Propy, a Silicon Valley-based real estate technology firm founded by Natalia

Karayaneva, spent years solving exactly that gap: how to make a blockchain transaction legally binding within existing property law frameworks. Their answer was to tokenize ownership through an LLC that held the physical deed, creating a bridge where the blockchain record and the legal record pointed to the same truth simultaneously.

In February 2022, that solution was tested in public for the first time. A five-bedroom home in Gulfport, Florida sold via blockchain auction for 210 Ether, approximately $653,000 at the time. A few months later, a Tampa condominium sold for $185,000 USDC through Propy's NFT Marketplace, with ownership transferring to the winning buyer's digital wallet within 15 minutes of confirmation. No paper. No wire transfer delays. And it held up legally.

Natalia Karayaneva reflected on the milestone: "Real estate has always been slow to innovate. But this moment showed that blockchain isn't just for digital assets. It's for the homes we live in."

Verifying Funds

Here's something that doesn't change no matter how sophisticated the system gets: every real estate transaction ultimately depends on trust in the money itself. In a blockchain-powered transaction, fund verification can happen instantly, transparently, and automatically. Every cryptocurrency or digital asset movement is recorded on the blockchain with an immutable timestamp and transaction history. Transactions can be looked up with a block explorer such as

Etherscan.io, providing a level of clarity that simply doesn't exist in traditional banking systems.

AI and Compliance: The New Gatekeepers

Companies like Merkle Science, Chainalysis, and Elliptic use AI-powered analytics and predictive risk modeling to detect, trace, and prevent illicit crypto activity before it enters the financial system. Real estate agents, brokers, and firms can now partner with these compliance platforms to verify funds in seconds rather than days.

Jen & Braeden's Digital Closing

Jen and Braeden had been saving for nearly a decade to buy their first home, a cozy ranch on a quiet tree-lined street, with a big backyard where their kids could play and a covered porch where they could drink coffee on Sunday mornings. When their offer was accepted, the excitement was electric.

But Jen and Braeden's transaction didn't unfold like the usual waiting game. Their closing ran on a blockchain-powered smart contract. When their offer was accepted, the purchase price, contingencies, and timelines were entered into the contract. Instantly, everyone involved, including their lender, the title company, the inspector, and even the county recorder, could view the same transparent set of terms in real time. Nothing hidden, nothing lost in translation.

On closing day, the smart contract checked each condition one last time. Financing: verified. Inspection: approved. Title: cleared. All green. With those triggers met, it automatically executed the transfer. A digital deed created, signed, and permanently stored, immutable, timestamped, and accessible to both parties within seconds.

Jen and Braeden sat in their living room that evening, the kids running in circles on the hardwood floors, amazed at how simple it all felt. The entire process, from offer to keys in hand, took less than a week. Their agent, Matt, later described it best: "It was like watching a closing that ran itself. Every step was transparent, every update automatic, and everyone knew exactly where the deal stood. No guesswork, just progress."

Why Blockchain Transactions Matter

Blockchain-powered transactions represent a fundamental redesign of how trust, time, and value move through the real estate system. By establishing a shared, immutable ledger, a single version of truth, every authorized participant can see, in real time, what stage a transaction is in and what conditions have been met.

Efficiency and Cost Savings

Smart contracts execute instantly once the criteria are fulfilled, on nights, weekends, and holidays alike. That efficiency translates directly into cost savings for agents, buyers, sellers, and lenders alike.

Trust and Transparency

Blockchain removes the opacity. No one can secretly alter records, delete documents, or manipulate timestamps. Buyers get confidence they truly own what they paid for; sellers get assurance that payment is verified before any transfer takes place.

Global Access

Blockchain opens the door to a borderless property market. An investor in New York could seamlessly close on a property in Dubai, with the entire transaction verified, executed, and recorded on-chain.

Companies Leading the Way

Public consensus will determine which platforms become standard over time, much the same way Carfax became the trusted source for vehicle history as dealers and consumers gradually aligned around a single, reliable tool.

Propy: Pioneering Blockchain Closings

Founded by Natalia Karayaneva, Propy's mission is to simplify and secure property transactions by moving them on-chain. Propy's blockchain-enabled deed recording system is now being tested by local governments in the U.S., including in Arizona, Florida, and Colorado.

Merkle Science: The Guardians of Financial Integrity

Merkle Science operates at the intersection of security, compliance, and analytics. Their AI-powered platform helps institutions, governments, and property technology companies detect suspicious activity, trace fund origins, and meet AML and CTF regulations.

Blocksquare: Tokenizing Real Estate for the Everyday Investor

Founded in Slovenia, Blocksquare built a blockchain infrastructure that allows real estate companies to tokenize physical properties, dividing ownership into digital shares. Other pioneers include Roofstock onChain, RealT, and Ubitquity.

Future Outlook: The 24/7 Real Estate Market

The closing table of the future won't be bound by time zones or office hours. Think about what happened when GPS navigation replaced paper maps. What once required expertise and careful advance planning can now be handled automatically by anyone with a phone. Real estate is heading the same direction. A buyer in Singapore will be able to invest in a Manhattan high rise with the same ease as someone buying their first home down the street.

Some fear this shift will replace professionals, but the opposite is true. The technology takes care of the transaction; the people take care of the relationship.

Key Takeaways

1. **Blockchain relocates trust, not just process.** The core shift is moving trust from institutions to mathematics. Smart contracts execute exactly as written, without interpretation, delay, or the possibility of quiet manipulation.
2. **Smart contracts are the mechanism.** They appear in every blockchain real estate application. Understanding them once means understanding the whole ecosystem.
3. **The first transactions have already happened.** Propy's 2022 blockchain sales proved the concept works within existing legal frameworks. The infrastructure is operational; adoption is now the limiting factor.
4. **Global access is real.** The property market is no longer limited by geography or currency.
5. **The 24/7 marketplace is the new standard.** Real estate will soon operate continuously, no banking hours, no borders, no downtime.

Take Action: Blockchain Transactions

- Explore Propy's platform (propy.com), free to browse. See what a blockchain-enabled closing dashboard looks like from an agent's perspective.
- Search "blockchain deed recording" plus your state to find out whether your county is running a pilot program. If not, that's a conversation worth starting with your local recorder's office.

- Ask your title company or escrow officer whether they've explored blockchain-integrated platforms. Their answer will tell you a lot about where your market is in the adoption curve.
- If you work with international buyers or sellers, research which cryptocurrencies and stablecoins are accepted by the platforms operating in your market. USDC is the most common starting point.

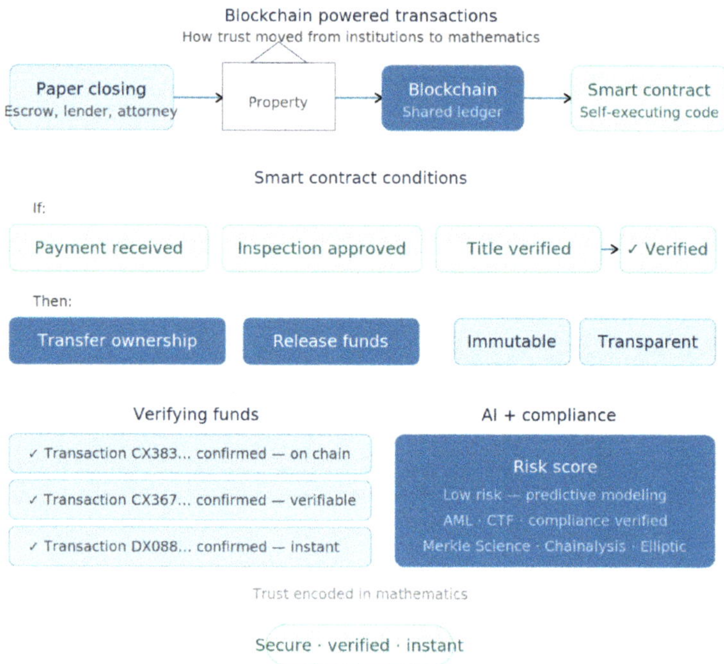

Chapter Two

Title & Legal Processes

Untangling property's paper chains

Every home, office building, and parcel of land sits on top of an invisible foundation: its title history. These records define who owns what, trace every transfer of ownership, and protect both buyers and sellers from fraud or disputes. In theory, this system forms the backbone of trust in the real estate industry. But in practice, it's messy, outdated, and alarmingly fragile.

Across the United States, property records are scattered among county clerk offices, title companies, legal archives, and government databases, each operating its own system with its own version of the truth. A missing signature, an unreleased lien, or an unclear boundary line can derail a closing, add costs in legal fees, or send buyers and sellers spiraling into weeks of delay.

A Case in Progress: New Jersey's Blockchain Revolution

In May 2025, Bergen County, New Jersey launched one of the most ambitious real-world blockchain title projects in the United States. The county signed a five-year agreement with technology firm

Balcony, built on Avalanche's blockchain infrastructure, to bring over 370,000 property deeds, representing nearly $240 billion in real estate value, onto the blockchain. The initiative is projected to reduce deed processing times by over 90 percent, cutting a process that currently takes up to 90 days down to a single day.

Here's what that looks like in practice for a Bergen County homebuyer. A title history that stretches back half a century, involving multiple owners, refinances, subdivisions, and lien releases, once requiring weeks of research through courthouse archives, will exist on a shared digital ledger accessible in seconds.

The benefits are already visible in Balcony's earlier work. During the company's rollout in the city of Orange, New Jersey, officials discovered nearly $1 million in previously uncollected revenue from discrepancies in tax assessments and missing filings, errors that had gone unnoticed until the blockchain ledger exposed them.

Understanding the Three Pillars of Title

Every real estate transaction begins and ends with one fundamental question: who owns the property, and who has the right to sell it? Traditionally, this process revolves around three core pillars: Ownership Records (the chain of title), Legal Conditions and Encumbrances (easements, liens, zoning), and Verification and Recording (lawyers, title companies, and county clerks confirming and recording the transfer). Today's title system remains slow, fragmented, and prone to human error.

The Hidden Cost of Title Errors

According to research from the American Land Title Association, more than a third of real estate transactions require some form of curative title work before closing, a figure that has remained remarkably consistent across decades of industry surveys. Every time a title company uncovers an inconsistency, the ripple effects are immediate: delayed closings, additional legal fees, renewed rate locks, and lost deals. Title insurers in the United States process billions of dollars in premiums annually. The U.S. property system still operates as a patchwork of more than 3,000 independent recording offices.

Lexi's Mystery Lien

Lexi had finally found her dream home, a sunlit, three-bedroom colonial with a fenced-in backyard perfect for her two children and their golden retriever. Her financing was lined up, the inspection passed, and everything seemed perfectly on track, until the title search returned with a single, cryptic entry: a lien from 1984.

It didn't make sense. The lien was supposedly filed against the property forty years earlier by a construction company that no longer existed. For weeks, Lexi's real estate agent and the seller's attorney traded phone calls with courthouse clerks sifting through microfilm and dusty archives. Lexi's rate lock ticked closer to expiration.

When the county finally produced the records, the truth was almost laughable. The lien had been satisfied and released the same year it was filed. The error had occurred during a clerical update in the 1990s: an entry logged twice, once correctly and once with no release tag. For nearly thirty years, no one had noticed. "I felt like I was being punished for someone else's mistake," she said later. "All that paperwork, all that waiting, and it was just a typo from forty years ago."

In a blockchain-based title registry, that scenario plays out differently. The lien and its release would have been recorded as a single, immutable entry on a shared ledger back in 1984. When Lexi's title company searched the blockchain record, the lien would have appeared with a clear status: Resolved. Verified. Released. A process that took five weeks could have taken five minutes.

From Microfilm to Blockchain: A Clerk's Perspective

When County Title Clerk Melissa Rossi first heard that they would be moving property records to a blockchain system, she was skeptical. She'd spent twenty years in the clerk's office, surrounded by walls of filing cabinets and the rhythmic hum of the microfilm scanner.

A few weeks after the system went live, a local attorney called searching for an old lien release from 1992. Melissa typed in the property's blockchain ID, and the entire record appeared, verified and clear, with the release linked directly to the lien. The attorney was stunned. "Wait, that's it?" he asked. Melissa laughed. "That's it. Apparently that's the blockchain technology they talk about." "We used to call them title chains because of all the paper you had

> *to pull together," she said. "Now we actually have a real chain: a blockchain. And it's the first one I've ever trusted not to break."*

Case Study: Georgia's Blockchain Land Registry

In 2016, the Republic of Georgia partnered with Bitfury to build one of the world's first blockchain-based land registries. To understand why that mission mattered, consider what property ownership looked like in Georgia before the project: in a country where land records were maintained in local offices, often on paper, a well-connected neighbor, a corrupt official, or a simple clerical error could quietly alter who owned what. Small farmers and rural landowners, people who had worked the same land for generations, had little recourse when disputes arose. The blockchain registry was designed precisely for them: a system where the record of ownership belonged to the ledger, not to whoever held the paper.

Within a few years, the initiative had recorded hundreds of thousands of property titles on the blockchain. Verification times dropped from several days to seconds, transaction costs fell sharply, and property disputes decreased. For the first time, a Georgian farmer could prove they owned their land not by producing a paper document, which could be lost, altered, or challenged, but by pointing to a cryptographic record that the entire network agreed was true. The uncertainty that had shadowed property ownership for generations just… stopped.

"The Georgian model has since inspired pilots in Sweden, Ukraine, Dubai, India, Japan, Canada, and across the United States."

Beyond the Deed: Your Home's Living Record

For most homeowners, the paperwork that defines their property lives in a binder somewhere, a drawer, a filing cabinet, maybe a folder they haven't opened since closing day. A roof warranty from 2019. A permit for the kitchen remodel. A receipt for the HVAC replacement. An inspection report from the previous owner. Individually, each document tells a small piece of the story. Together, they tell the full history of a home. The problem is that together, they almost never are. Blockchain changes that. The same immutable ledger that records a deed transfer can record every meaningful event in a home's life, repairs, renovations, permits, warranties, appliance installations, and inspections, each timestamped, verified, and permanently linked to the property itself. Not stored in a binder that gets lost in a move. Not buried in an email thread. On-chain, attached to the address, accessible to anyone with the right to see it.

For buyers, this is transformative. Instead of relying on a seller's disclosure form, which is only as accurate as the seller's memory and honesty, a buyer could pull up a home's complete event history in seconds. When was the roof last replaced? What contractors worked on the electrical system, and were they licensed? Is that water heater still under warranty? The blockchain doesn't forget, and it can't be edited after the fact.

For homeowners, it's a living record that builds real value. A well-documented home, one with a verified chain of maintenance, permitted upgrades, and transferable warranties, is a more trustworthy asset at resale. Think of it less like a filing cabinet and more like a Carfax report for your home: instant, accessible, and verifiable by anyone in the transaction.

The deed was always just the beginning of a property's story. Blockchain is finally giving us a way to record the rest of it.

Key Takeaways

6. **Title fraud and error cost the industry billions annually.** More than a third of real estate transactions require curative work before closing, a problem blockchain addresses at the architectural level, not through insurance patches.
7. **Blockchain makes title history instantly verifiable.** Every lien, payoff, and ownership transfer stored on-chain is permanent, tamper-proof, and accessible in seconds rather than weeks.
8. **The transition is underway.** Bergen County, New Jersey's 370,000-deed blockchain initiative is the largest of its kind in the U.S., and it's already in progress.

9. **Smart contracts automate the closing sequence.** As introduced in Chapter 1, smart contracts eliminate the manual handoffs between escrow, title, and recording.

Take Action: Title & Legal Processes

- Look up your county's property records portal and test how far back you can trace a title history online.
- If you're a title professional or attorney: research Propy's pilot programs in Arizona, Florida, and Colorado.
- Search for "Balcony blockchain" and "Avalanche land registry" to follow the Bergen County initiative.
- Ask your title insurance provider whether they're exploring blockchain-integrated underwriting.

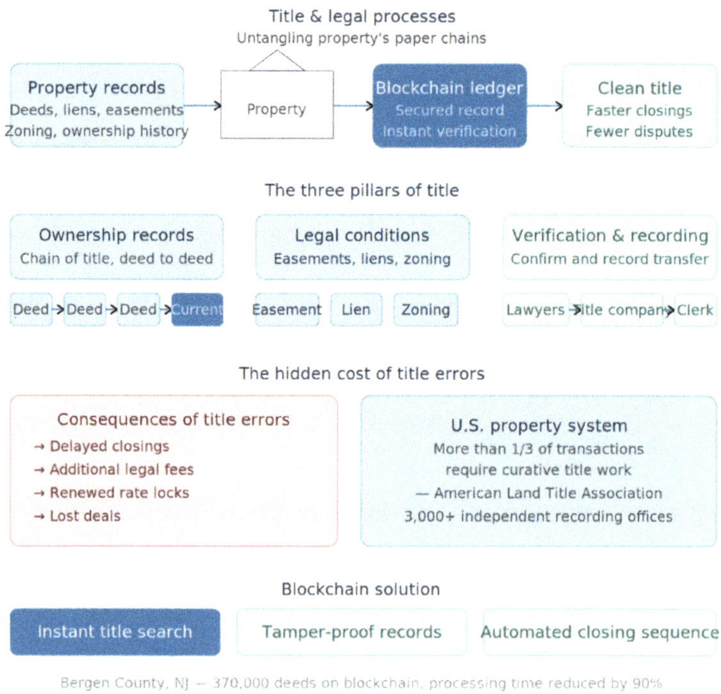

Chapter Three

Tokenized Real Estate Assets

Opening ownership to everyone

The Barriers to Real Estate Investment

There's a teacher in North Carolina who has spent twenty years watching the apartment building on her street change hands. First a local owner, then a regional investor, then a private equity firm that raised the rents by 40 percent in three years. The building generates real wealth every single month. She walks past it on the way to work. She will never own a share of it.

That gap between the people who live in proximity to wealth-generating assets and the people who actually own them has defined real estate investing for centuries. Tokenization is the most direct technological challenge to that exclusivity that has ever existed.

Real Estate Tokenization

At its core, real estate tokenization is the process of transforming the ownership of physical property into digital form, breaking it down into secure, traceable, and tradable pieces on a blockchain.

The process begins by placing a property into a legal entity, such as a Limited Liability Company (LLC) or a Special Purpose Vehicle (SPV). The ownership of that entity is then divided into digital tokens, each representing a fractional share of the underlying asset.

This enables fractional ownership, liquidity through tradeable tokens on compliant marketplaces, and programmable rights where each token carries built-in logic defining entitlements. Instead of requiring millions, investors can participate with hundreds.

From Crowdfunding to Blockchain

Before blockchain entered the scene, real estate crowdfunding was the first major attempt to open property investment to the everyday investor. Platforms like Fundrise, RealtyMogul, and CrowdStreet allowed small investors to pool money online. But once you invested, your money was locked up for years with no secondary market to exit.

Blockchain changed that equation entirely. Where crowdfunding created access, blockchain created autonomy. Smart contract automation handles income distribution, ownership transfers, and governance votes automatically. From startups like Lofty.ai to enterprise-grade solutions like LiquidFi and Blocksquare, the tools to trade real estate digitally are live today.

My Road Trip Investment in Tokenized Real Estate

It was a cold December afternoon in 2024, somewhere along a long stretch of highway in the Midwest, the kind of drive where your mind drifts between playlists, podcasts, and possibility. I was in the passenger seat, phone in hand, scrolling through Lofty.ai, a tokenized real estate platform I'd been researching for weeks.

With just $50, I became a fractional owner of a rental property, a real physical home, through a tokenized real estate platform powered by blockchain. The process was shockingly seamless. I set up an Algorand-compatible wallet, completed identity verification through the platform's KYC system in a matter of seconds, and made my investment in USDC. Within minutes, the smart contract had executed, my tokens were secured, and my share of ownership was live.

As fate would have it, our route home passed directly through the small Michigan town where this property stood. I pulled up the address, and as we drove past quiet streets dusted in winter snow, I spotted the very building whose income I now partially owned.

I laughed out loud in the car. Not at anything in particular. Just at how weird and obvious it felt at the same time. I hadn't invested in a vague promise or a digital illusion. I'd invested in a real house, on a real street, with real tenants, governed by blockchain, secured through smart contracts. A property hundreds of miles away, bought in minutes from the passenger seat. And somewhere on that quiet snowy street, a teacher might have been *looking out her window at the same kind of building, one that, with $50 and a phone, she could now own a piece of too.*

How Tokenized Real Estate Works Behind the Scenes

1. Legal Structures

In tokenized real estate, you don't technically own the building itself. You own a piece of the legal entity that does. Most platforms use a Special Purpose Vehicle (SPV), often structured as an LLC, that holds the title to the property. This legal setup keeps everything compliant with existing real estate and securities laws.

2. Smart Contracts

As introduced in Chapter 1, smart contracts automate what once required entire departments of accountants, lawyers, and escrow agents. When rent is collected, the smart contract automatically distributes payments to each token holder based on their ownership percentage.

3. Secondary Markets

Property stakes can be listed on regulated secondary markets where investors buy and sell digital shares instantly, similar to trading stocks. This introduces liquidity to one of the most illiquid asset classes on Earth.

4. Compliance and Taxation

Tokenized real estate still operates within familiar financial rules. The IRS classifies digital assets as property, meaning they are subject to capital gains and income tax laws. Given the complexity of cost-basis tracking across fractional tokens, consulting a qualified tax professional for your specific situation is always advisable.

Case Study: Lofty.ai's Governance Model

The road trip investment earlier in this chapter describes what buying into a tokenized property feels like. What it doesn't describe is what happens after the purchase. Specifically, what it means for ordinary investors to have a governance stake in a real property.

- Voting Rights: Each token holder gets a proportional voice in key property decisions, such as approving repairs, adjusting rent, or replacing management.
- Income Distribution: Rental profits are disbursed automatically to wallets via blockchain, without middlemen or delays.
- Transparency: Every property's performance, governance vote, and transaction history is visible to token holders at any time.

To maintain fairness, no single investor can own more than 15% of a property's tokens. In practice, a group of investors spread across multiple states can collectively vote to replace an underperforming property manager, approve a repair budget, or authorize a rent increase. Those votes happen on-chain, the results are recorded permanently, and the outcome executes automatically.

It's the same apartment building on the teacher's street, but now the door is open, and the people inside have a voice in how it's run.

Case Study: LiquidFi & Blocksquare

LiquidFi: The Rails

If Lofty.ai is the platform that puts fractional ownership in the hands of individual investors, LiquidFi is the infrastructure layer that makes institutional-grade tokenization possible: the rails, not the train. LiquidFi's documented work includes reducing mortgage-backed securities reporting time from 55 days to 30 minutes on the Stellar blockchain. The result is a tokenized asset that can survive the due diligence process of a bank, a pension fund, or a government agency, not just a retail investor browsing a marketplace at 11pm.

Blocksquare: The Train

Blocksquare is the engine that lets a property developer in Berlin, a commercial landlord in Singapore, or a real estate firm in Chicago launch their own tokenized offering on their own website, powered by Blocksquare's infrastructure running quietly in the background. The technology is now deployed in 29 countries. Together, LiquidFi and Blocksquare represent two distinct entry points into the same ecosystem: one building the standard, the other distributing it.

Did You Know?

- *Blocksquare has tokenized over $200 million in real estate across 29 countries, with an 800% increase in total value locked across the broader tokenized real estate sector in 2025 alone.*
- *LiquidFi reduced mortgage-backed securities reporting time from 55 days to 30 minutes using blockchain.*
- *Deloitte projects the tokenized real estate market to grow from $300 billion in 2024 to over $4 trillion by 2035.*
- *In mid-2025, Dubai tokenized two luxury properties that sold out within minutes to buyers from over 35 countries, with roughly 70% of those buyers being first-time investors in Dubai real estate.*
- *The US, EU, Singapore, and UAE all now have live regulatory frameworks for tokenized real estate. The question is no longer whether it's legal, but which framework fits the deal.*

Future Outlook: The Global Real Estate Exchange

If tokenization represents the bridge, the destination is already coming into view: a global real estate exchange where property trades with the same fluidity as stocks. Institutional players are already watching closely. Pension funds, sovereign wealth managers, and asset management firms see tokenization as the next frontier of diversification. The teacher in North Carolina doesn't have to watch from the sidewalk anymore.

Key Takeaways

10. **Tokenization is the most direct challenge to real estate's exclusivity.** It removes the capital and geographic barriers that have historically kept ordinary investors out.

11. **The platforms are live and operational.** Lofty.ai, LiquidFi, and Blocksquare have proven that fractional property ownership can be fast, compliant, and secure at real scale.

12. **Smart contracts automate the landlord.** Income distribution, governance votes, and ownership transfers all happen automatically.

13. **Regulation is catching up, not blocking.** Tokenization works within existing legal frameworks. The pathway is clear; the infrastructure is ready.

14. **A global real estate exchange is coming.** The technology works. The platforms exist. The laws are catching up. What's left is participation.

Take Action: Tokenized Real Estate

- Create a free account on Lofty.ai and browse available properties before investing anything.

- Make a small test investment of $50 to $100 in a tokenized property. The goal isn't financial return; it's firsthand experience of the process, from wallet setup to income distribution.
- Research how Blocksquare's white-label model works if you're a developer or property manager.
- If you work with high-net-worth investors, start a conversation about tokenized real estate as part of a diversified portfolio.

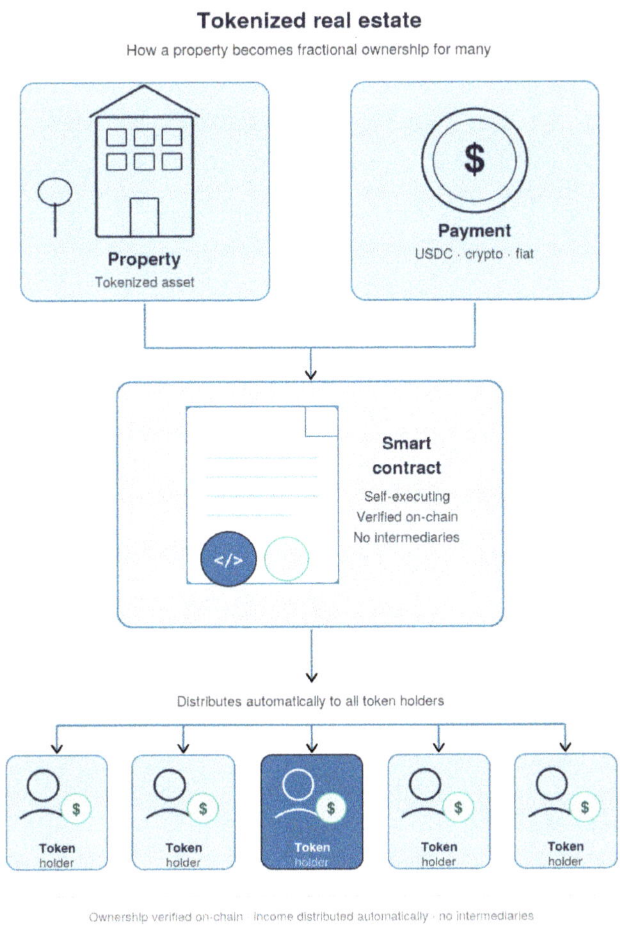

Chapter Four

Metaverse & Virtual Land

A new frontier, with eyes wide open

Every generation redefines what makes a location valuable. Rivers gave way to railroads, railroads gave way to highways, highways gave way to broadband. Each shift rewrote the map of where people wanted to be. The metaverse is the next redrawing, except this time, the map isn't physical at all.

Between 2021 and 2022, companies like Nike, Gucci, and Samsung began acquiring and developing land in virtual platforms like The Sandbox and Decentraland. Behind all of it, blockchain technology provided the same fundamental traits as physical real estate: scarcity, verifiable ownership, and tradable value. Whether that experiment succeeds in the long run is still an open question. But the question itself is not going away.

Market Reality

Before diving into the potential of virtual real estate, intellectual honesty requires acknowledging what happened after the 2021 to 2022 peak. Metaverse land values experienced a severe correction in the years that followed. According to a CoinGecko study, average metaverse land prices had fallen roughly 72 percent from their highs by mid-2024, with The Sandbox down approximately 95 percent, Decentraland down approximately 89 percent, and Otherside down approximately 85 percent.

Both things are true: the technology is real, and the 2021 prices were not. This chapter tries to keep both in view.

Why the Metaverse Still Matters for Real Estate

The 2022 crash didn't kill the question. It just stripped the hype off it. For real estate professionals, the right question isn't really "is the metaverse valuable?" It's "what does it mean for how value works?" If digital spaces become persistent, inhabited, and commercially active, even at a fraction of the scale the 2021 hype suggested, then the principles of real estate still apply. Location. Scarcity. Development. Leasing. The medium changes; the economics stay pretty recognizable.

What Is the Metaverse?

The metaverse is not a single place or a single platform. It's a collective term for immersive, interconnected virtual environments where users interact as avatars, conduct commerce, attend events, and increasingly, work. Blockchain gives these worlds permanence and structure, ensuring that digital assets are owned by you, not by the platform that hosts them.

In the metaverse, ownership lives on the blockchain. Each parcel of land can be represented as a non-fungible token (NFT), a cryptographic proof of ownership stored on a decentralized ledger. Virtual real estate is scarce by design: Decentraland permanently capped its world at 90,601 parcels.

Metaverse Real Estate in Action

Case Study: Decentraland's Land Rush

The Decentraland Fashion District sale is one of the most instructive case studies in virtual real estate, not because of the $2.4 million price tag, but because of what happened next. In November 2021, Metaverse Group purchased a single parcel for $2.4 million USD. Within a year, brands like Dolce & Gabbana, Estée Lauder, and Forever 21 were participating in Metaverse Fashion Week inside that space. By any reasonable measure, the vision worked: the events happened, the brands showed up, the platform functioned. And yet, by 2024, that same parcel was worth a small fraction of what Metaverse Group paid for it. The lesson isn't that virtual real estate doesn't work. It's that the price of a digital asset and the utility of a digital asset are two different things, and in 2021, the market confused them catastrophically.

Case Study: Republic Realm & The Sandbox

Republic Realm's purchase of a 576-parcel estate in The Sandbox for approximately $4.3 million in November 2021, the largest single virtual land transaction at the time, tells a subtler version of the same story. Republic Realm wasn't a crypto enthusiast making a speculative bet. It was a real estate investment firm that figured virtual land next to major brand activations would appreciate the same way physical land next to major commercial districts does. That's not crazy reasoning. The part they didn't fully account for was that the medium itself hadn't yet proven it could hold the user engagement that makes location valuable in the first place. Scarcity without sustained demand isn't a real estate asset. It's a lottery ticket.

Case Study: Roblox & The Next Generation

The most underappreciated case study in this chapter has nothing to do with blockchain. Roblox has spent two decades running a behavioral experiment in digital ownership psychology, and the results are pretty clear. With approximately 88 million daily active users as of Q3 2024, Roblox players create virtual environments and assign genuine scarcity value to digital items without any blockchain infrastructure whatsoever. The generation currently entering adulthood doesn't need to be convinced that digital ownership is real. They grew up knowing it. The only question is whether the real estate industry will have its act together by the time they show up as buyers.

Historical Context: From Second Life to Blockchain Ownership

Back in 2003, a platform called Second Life gave users unprecedented creative freedom. But there was one fundamental flaw: everything inside Second Life belonged to the company that owned the servers. If the servers went offline, digital empires vanished with them. Blockchain technology fixed that. It introduced decentralized ownership, transforming virtual property from a line of code in a company database into a cryptographically secured, publicly verifiable asset. Second Life showed us what digital societies could look like. Blockchain showed us how to own them. eXp Realty, one of the largest brokerages in the world by agent count, is a real-world example of a traditional real estate company that has embraced virtual world operations. Its agents collaborate, train, and conduct business inside eXp World, a virtual campus powered by FrameVR, a browser-based platform accessible from any device. For eXp, the virtual world isn't a novelty. It's part of how they work.

The Next Generation of Property Owners

For the next generation, this isn't foreign. It's familiar. Today's kids and teenagers already spend significant hours each week in immersive environments like Roblox, Minecraft, and Fortnite Creative (Fortnite's player-built world mode, where users design and inhabit their own virtual spaces). They build virtual homes, customize avatars, and trade digital assets with the same enthusiasm past generations had for baseball cards or model cars. When these users grow up and enter the workforce, buying digital property or attending a virtual open house won't feel novel; it will feel obvious.

When the Metaverse Is a Mirror: Digital Twins and the Question of Who Owns Your Plot

Most metaverse platforms invent their geography. Decentraland, The Sandbox, Otherside, their maps are imagined worlds, unconnected to any physical address. But a different and more provocative idea is already being built: a metaverse that is a precise digital twin of the real world, where every plot of virtual land corresponds exactly to a real location on Earth.

SuperWorld is one of the earliest attempts at this. The platform maps the entire surface of the Earth into 64.8 billion unique plots, each covering a defined area of latitude and longitude, with every plot recorded as an NFT on the Polygon blockchain. Own a plot in SuperWorld, and you hold the digital property rights to that coordinate on the platform. The concept is genuinely compelling. The execution, however, has struggled to find its audience. SuperWorld has remained a niche product, never crossing into mainstream adoption, held back by the same friction that has slowed the broader metaverse: not enough people using it to make ownership feel meaningful, and not enough infrastructure to make the experience feel real.

But the idea itself is not the problem. It is simply ahead of the technology and the public that needs to be ready for it. When a more widely adopted, better developed version of this concept arrives, and the trajectory of AR hardware, spatial computing, and blockchain infrastructure suggests it will, the possibilities it unlocks are significant. In that world, owning a plot means you can build on it, decorate it, host experiences, run advertising, and target audiences based on real-world foot traffic at that exact GPS coordinate, all governed automatically by smart contracts, all recorded immutably on chain.

Which raises a question that the real estate industry hasn't fully reckoned with yet: what happens when someone else already owns the digital plot that sits on top of your physical property?

In the physical world, your deed defines what you own. In a digital twin metaverse, that deed means nothing unless the platform recognizes it. A homeowner in Chicago could discover that a virtual investor in Tokyo purchased their digital plot years ago and has been monetizing the airspace above their address ever since. There is currently no legal framework that connects physical property ownership to its digital twin equivalent. The blockchain verifies who owns the token. It does not ask whether that person owns the land underneath it.

This is where blockchain's role becomes both powerful and unresolved. The same technology that can make ownership transparent and tamper-proof could, in theory, create a parallel property rights system that either complements or conflicts with the physical one. The concept isn't far removed from air rights or broadcast frequencies. Society has always found ways to layer new forms of ownership on top of the physical world. Digital twin property rights may simply be the next iteration.

What's clear is that property owners of the future will want to own all aspects of their address, physical, digital, and virtual. The deed was always just one layer of what it means to own a place. The layers are multiplying.

Key Takeaways

15. **The technology is real; the 2021 speculation was not the same thing.** Blockchain-verified virtual ownership is a genuine innovation. The speculative cycle that surrounded it does not diminish the underlying infrastructure.

16. **Virtual real estate follows recognizable economics.** Scarcity, location, development, and leasing all apply. The medium has changed; the principles have not.

17. **Market reality demands perspective.** The 2021 to 2022 peak was followed by declines of 85 to 95 percent across major platforms. Long-term value will depend on genuine user adoption.

18. **The next generation is already there.** Young people growing up in Roblox and Minecraft are tomorrow's buyers and investors. To them, owning a virtual asset will feel as natural as owning a physical one.

Take Action: Metaverse & Virtual Land

- Create a free Decentraland or The Sandbox account and spend 30 minutes walking the virtual environment.
- Review current floor prices for virtual land parcels on platforms like OpenSea. Compare them to 2021 peaks.

- If you work with younger clients (under 35), ask them about their relationship with virtual environments and digital assets.
- Follow one metaverse platform's governance forum (Decentraland's DAO is publicly accessible) for 90 days.

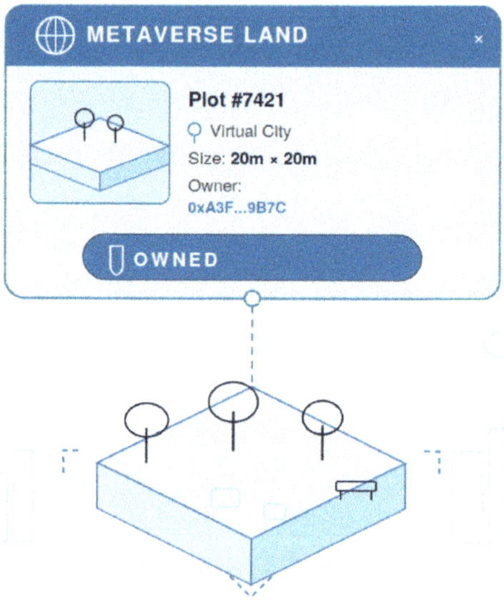

Each plot is a unique NFT · ownership recorded immutably on-chain · tradeable on secondary markets

Chapter Five

Digital Land Overlays on Real-World Space

The invisible layer above every building

Kalysta ran a cozy café tucked between high-rises in downtown Chicago, the kind of place where the morning rush never stopped and every customer seemed to have their favorite seat. Business was steady, but competition was fierce. Within a three-block radius, there were twelve other coffee shops. One afternoon, a representative from a local tech startup stopped by for an espresso and left her with an idea that would change the way she thought about her business forever.

"We're building a network of digital advertising overlays," he explained. "What if your café had ad space in the air above it, visible through AR glasses or phones? No construction, no hardware, no clutter. Just digital projection rights tied to your building's coordinates. You'd earn monthly income from the digital layer above your shop."

Kalysta was intrigued. Within months, tourists wearing AR glasses saw whimsical, animated coffee beans floating above the café. Every campaign was temporary, automatically launched and removed through smart contracts. Payments flowed into her digital wallet without paperwork, invoices, or negotiation delays. Her shop saw a 20 percent increase in foot traffic within three months. Kalysta had entered a new kind of real estate business, one where the invisible layer above her roof was worth as much as the tables inside her café.

Kalysta's story is illustrative. The technology she's using doesn't exist at commercial scale yet. But every component of it does. The question isn't whether Kalysta's world arrives. It's whether real estate professionals will be ready when it does.

For centuries, property rights have revolved around what could be touched, built, or seen. Most property owners today have no claim to the virtual layers floating above or surrounding their real estate. That unclaimed layer represents one of the largest frontiers of unrealized value in modern real estate.

Who Owns the Invisible Layer Above Our Heads?

As soon as digital overlays hold value, ownership becomes critical. A blockchain-verified record can assign, track, and enforce rights to a specific digital coordinate, ensuring the same parcel of virtual airspace can't be sold twice. When Kalysta leases her digital layer to an advertiser, a smart contract records the transaction, releases the payment, and updates the ledger instantly, all automatic, all auditable, and all tamper-proof.

How Digital Overlays Work

A digital land overlay is a virtual layer mapped onto real-world geography, built on three foundational technologies. GPS and Geospatial Mapping provide precise coordinates, anchoring every digital object to a real-world location. Blockchain serves as the deed system, recording who owns each digital parcel. Augmented and Virtual Reality (AR/VR) bring these overlays to life, allowing people to see, interact with, and experience them through their devices or wearable smart glasses.

Advertising & Branding

A downtown broker claims the AR rights above her firm's building. Through smart glasses, passersby see a 3D banner gently rotating above the entrance. A developer launching a luxury condo projects a digital tour that soars the viewer from the sidewalk up to the penthouse.

Tourism & Culture

Historic neighborhoods become interactive time capsules. As visitors stroll past restored homes, digital plaques hover in midair showing photos of what stood there a century ago.

Commerce & Retail

A shopping center layers its storefronts with virtual sales tags and animated window displays that adapt in real time. Residential agents host curbside previews where a buyer standing outside a listing can view 3D floor plans, price history, or mortgage calculators projected right into the air in front of the home

Owning the Digital Dimension

In many ways, digital overlay rights are the 21st-century version of airspace rights. Just as property owners can sell or lease the air above their buildings for signs, antennas, or billboards, they can now sell or lease the digital space above. Global outdoor advertising is a multi-billion-dollar industry, and AR overlays could multiply that reach dramatically, without ever printing or mounting a single sign.

Historical Context: From Air Rights to Digital Rights

In the early 20th century, as cities began to rise skyward, lawmakers faced a new question: who owns the air above a building? The answer gave birth to air rights, the legal ability for property owners to control, sell, or lease the vertical space above their land. Many of Manhattan's most iconic towers exist because developers purchased the invisible space above their neighbors.

Now we're witnessing the next evolution of that same logic: digital rights. Where air rights allowed cities to build upward, digital overlay rights allow them to build beyond. Legally, the idea of owning intangible property isn't new. We already recognize intellectual property, broadcasting frequencies, and satellite orbits as valuable real estate in the digital domain. Digital overlays follow the same principle.

Emerging Concepts and Possibilities

1. Global AR Mapping

Tech innovators are racing to create persistent, shared 3D maps of the real world, known as spatial anchors. These maps allow digital objects to stick to real-world locations with GPS-level precision. The next logical step will be property-linked overlays that transform cities into multi-layered, interactive environments.

2. Blockchain-Verified Geography

Every coordinate, image, or 3D object linked to a location can be recorded on-chain, time-stamped, authenticated, and tied to an owner. This system would ensure no duplicate claims to the same digital airspace and permanent documentation of who created, owned, or modified each overlay: the deed registry for the digital world.

3. The Spatial Web

As computing power shifts from screens to environments, we're entering what futurists call the Spatial Web, a version of the internet mapped directly onto physical space. A local bakery's website could appear as a floating menu outside its door. A city government's services could hover over public buildings, accessible through AR glasses.

4. Residential and Private Displays

Digital overlays won't be limited to commercial use. Homeowners and small business owners will be able to use the layer above their own property for personal expression, seasonal decoration, and grassroots promotion. Think of a home lit up with animated holiday displays visible through AR glasses, or a backyard greenhouse advertising fresh produce to neighbors walking past, or an Etsy shop owner projecting their storefront into the air above their front door. The same tools that let a corporation run a billboard campaign could let an individual personalize their corner of the digital world. As with any new property right, HOA regulations governing the content, brightness, and timing of residential digital displays are sure to follow.

Future Outlook: Smart Cities with Digital Zoning

As AR adoption accelerates and blockchain infrastructure scales, smart cities will begin managing not just traffic and utilities, but digital airspace. Cities may develop digital zoning laws that regulate virtual signage density, brightness levels, or animation timing.

In the not-too-distant future, a standard MLS listing may look familiar but carry one new line: Digital Overlay Rights. Appraisers will review on-chain records of who has leased the airspace above a property. A building with consistent digital income will earn a higher valuation. The physical city is finite. The digital city being built on top of it is effectively unlimited. Kalysta already knows that. The question is when the rest of the industry catches up.

Key Takeaways

19. **Digital space above physical property is the next unclaimed asset class.** Most property owners today have no claim to the AR layer above their buildings. That is about to change.

20. **Blockchain brings order to the invisible layer.** Every overlay can be recorded on an immutable ledger, preventing duplicate claims, enforcing leases, and automating payments.

21. **The historical precedent is solid.** Air rights, subterranean rights, broadcast frequencies. Society has always found ways to own and monetize space beyond the physical. Digital overlay rights are the next iteration.

22. **Real estate becomes an experience layer.** Tomorrow's listings will include overlay rights alongside square footage and lot size.

Take Action: Digital Land Overlays

- Download a consumer AR app (Snapchat Lens Studio, Niantic's Lightship, or similar) and experiment with placing a digital object at a real-world GPS coordinate.
- Research how your city or municipality currently handles signage and billboard permits. The same regulatory framework will eventually extend to AR overlays.
- If you manage commercial properties, document the "digital airspace" above your buildings: foot traffic patterns, adjacency to landmarks, event traffic.
- Follow AR development news from Apple (Vision Pro), Meta (Quest), and Niantic for 90 days.

Digital overlays visible through AR glasses · governed by smart contract · payments automated on-chain

Conclusion

Building the Future, Block by Block

What's Changing?

If you've read this far, you've traveled through five different corners of what blockchain is doing to real estate. Faster transactions. Tamper-proof title records. Fractional ownership. Virtual land. Digital airspace. Each chapter describes a different application. But stepping back, they all describe the same underlying shift.

> *"Blockchain doesn't add a new layer to real estate. It relocates trust."*

For centuries, real estate transactions worked because we trusted institutions: banks to hold escrow, county offices to maintain title records, escrow agents to release funds at the right moment, lawyers to interpret and enforce agreements. That trust was earned, mostly, but it was expensive: in time, in friction, and in all the people you had to pay to move trust from one hand to the next.

Blockchain moves trust into mathematics instead. A smart contract doesn't trust anyone. It verifies. A blockchain title registry doesn't trust the county database. It replaces it with something that can't be quietly corrupted. That shift, from institutions to mathematics and from intermediaries to code, is the actual revolution. Everything else is just what that revolution looks like in practice.

The People in These Pages

Technology alone doesn't shape the future. People do. Jen and Braeden experienced what happens when a closing is governed by code rather than coordination. Lexi saw what forty years of paper-based recordkeeping can cost a single homebuyer. Melissa Rossi spent two decades managing the daily grind of a system that was never quite good enough before watching it actually get fixed. The road trip investor put $50 into a rental property from a car on a highway, then drove past the building it represented. Kalysta found value in the invisible layer above her café that had always existed but never had a system of ownership.

Each of these is a story about trust being relocated. From uncertainty to verification. From exclusion to access. From opacity to legibility. That's what this technology actually does to the people inside it.

What This Means for Real Estate Professionals

Every agent, broker, developer, investor, and entrepreneur now stands at the same crossroads. The mechanics of real estate are being rebuilt from the foundation up. The professionals who thrive in that environment won't necessarily be the ones who master every blockchain protocol. They'll be the ones who understand what's actually changing: where trust lives, how ownership is verified, and what that means for the value they deliver to clients.

Knowing your way around blockchain will become about as basic to real estate practice as understanding mortgages, deeds, or zoning laws. Not because every agent needs to audit smart contract code, but because clients are going to ask, and the professionals who can actually explain what's happening will earn the trust that matters most. No smart contract can sit across from a nervous first-time buyer and tell them everything is going to be fine. No algorithm can read the room when a deal starts to fall apart. No blockchain can replace the agent who shows up, listens, and guides someone through one of the biggest decisions of their life. The trust that technology builds is mathematical. The trust that people build is personal. Both matter. But only one of them is irreplaceable.

What Comes Next

In the coming decade, global blockchain property registries may replace fragmented local databases. Hybrid property rights will bundle physical deeds with digital overlays. A 24/7 real estate market will operate across time zones and currencies. Smart cities powered by the convergence of blockchain, AR, IoT, and AI will emerge where buildings communicate, contracts self-execute, and property data is instantly available.

The future of real estate isn't arriving someday. It's being built now, block by block, in county recorder offices in New Jersey, on blockchain platforms in Michigan, in virtual fashion districts in Decentraland, and in the digital airspace above a coffee shop in Chicago where a smart contract just executed a lease for space that didn't exist until someone thought to own it.

Final Thoughts

If there's one skill this moment demands, it's the willingness to ask what's actually changing, not just what's new. New tools appear every week. What changes slowly, and what matters permanently, is where trust lives and how ownership works. Blockchain is shifting both.

The next foundation we lay won't be made of concrete, but of code and collaboration, built by people who understood what was changing before everyone else caught up. You've read the book. You know what's changing. What you do with that knowledge is the part that matters.

Where to Start

A Practical Guide for Real Estate Professionals

The action steps at the end of each chapter are organized here by role. Pick the track that matches your current position. None of these require technical expertise. They require curiosity and fifteen minutes of focused attention.

If You're a Real Estate Agent or Broker

- Create a free account on Propy (propy.com) and walk through the platform as if you were listing a property.
- Search "blockchain deed recording" plus your county or state. If a pilot program exists, attend any public meetings about it.
- Create a free account on Decentraland or The Sandbox and spend 30 minutes exploring. When clients ask about virtual real estate, and younger clients increasingly do, firsthand experience is worth more than any summary you can read.
- Document the "digital airspace" around your listings: foot traffic, proximity to landmarks, event density. These will be appraisal metrics within a decade.

If You're a Real Estate Investor

- Create a digital wallet (Metamask or Coinbase Wallet are the most accessible starting points) and make a small test investment on Lofty.ai. The process of going from zero to fractional property owner takes about 20 minutes and $50.

- Compare the current floor prices of Decentraland and Sandbox parcels against their 2021 peaks on OpenSea.
- If you have an existing portfolio, ask your attorney whether any of your LLCs could be structured as tokenization-ready entities.
- Follow the Bergen County, New Jersey blockchain initiative. Its outcomes will set precedent for every county that follows.

If You're a Developer or Property Manager

- Research Blocksquare's white-label model (blocksquare.io). If you have a project coming to market in the next 24 months, understanding whether tokenized capital raising is viable is worth the investigation now.
- Talk to your title company about blockchain-integrated underwriting. Ask specifically whether they're exploring platforms that reduce manual curative work.
- If you manage commercial properties, build a simple spreadsheet tracking foot traffic patterns, proximity to landmarks, and seasonal event density for each asset.
- Read LiquidFi's documentation on institutional tokenization (liquidfi.com).

If You're a Title Professional or Real Estate Attorney

- Review Propy's pilot programs in Arizona, Florida, and Colorado.
- Read at least one smart contract template from a tokenized real estate platform. You need to understand what the contract is designed to accomplish and what the legal exposure points are.
- Research how your state bar or title association is addressing blockchain deed recording.
- The ALTA Annual Conference, Inman Connect, and the Blockchain Real Estate Summit all regularly feature blockchain real estate programming.

If You're an Individual Investor or Curious Reader

- Start with the glossary at the back of this book. Make sure you can explain blockchain, smart contract, tokenization, NFT, and on-chain in plain language.
- Follow one blockchain real estate company (Propy, Lofty.ai, or Blocksquare) on LinkedIn for 90 days. For an inside look at how a tokenized real estate platform is built in real time, follow Jerry Chu, co-founder of Lofty.ai, on LinkedIn. He shares a candid, building-in-public perspective on the journey, the challenges, and where the industry is heading, the kind of transparent communication that is rare at any level of the industry.
- Make one small tokenized investment (under $100) on a platform like Lofty.ai. The goal is experience, not return.
- Share what you're learning. The conversations you have about blockchain real estate will sharpen your thinking faster than any amount of additional reading.

The topics covered in The Future is Verified involve financial actions. Nothing in this book constitutes financial advice. Consult a qualified financial advisor before making any investment decisions.

Glossary

Key Terms

The following terms appear throughout this book. Definitions are written for a general audience and prioritize practical understanding over technical precision.

AML (Anti-Money Laundering): Laws and processes requiring financial institutions and real estate professionals to verify that funds used in transactions are not derived from illegal activity.

AR (Augmented Reality): Technology that overlays digital information or images onto the physical world, typically viewed through a smartphone camera or smart glasses.

Blockchain: A distributed digital ledger that records transactions across a network of computers. Once data is recorded, it cannot be altered without changing all subsequent records and gaining network consensus, making it tamper-resistant.

Block Explorer: A publicly accessible tool that allows anyone to view the history and details of transactions recorded on a blockchain. Etherscan.io is an example for the Ethereum network.

Cryptocurrency: A digital or virtual currency secured by cryptography and typically operating on a blockchain. Examples include Bitcoin (BTC), Ether (ETH), and USDC.

CTF (Counter-Terrorism Financing): Regulations designed to prevent financial systems from being used to fund terrorist activity, often enforced alongside AML requirements.

DAO (Decentralized Autonomous Organization): An organization governed by rules encoded as smart contracts on a blockchain, where members vote on decisions proportional to their token holdings.

DeFi (Decentralized Finance): Financial services and products built on blockchain networks that operate without traditional intermediaries such as banks or brokerages.

Digital Twin: A virtual replica of a physical property or object, created using 3D modeling or AR/VR technology.

Fractional Ownership: A model in which multiple parties each own a percentage of a single asset. In tokenized real estate, each token represents a fractional share of a property.

KYC (Know Your Customer): The process by which a business verifies the identity of its clients, typically to comply with AML and other financial regulations.

LLC (Limited Liability Company): A legal business structure that provides its owners with limited personal liability. In tokenized real estate, an LLC often holds the title to a property.

Metaverse: A collective term for immersive, interconnected virtual environments where users interact as avatars. Examples include Decentraland, The Sandbox, and Otherside.

NFT (Non-Fungible Token): A unique digital asset recorded on a blockchain that certifies ownership of a specific item. Unlike cryptocurrencies, each NFT is distinct and not interchangeable.

On-Chain: Data or transactions that are recorded directly on a blockchain and are therefore publicly verifiable, tamper-resistant, and permanent.

RWA (Real-World Asset): A physical or financial asset represented as a token on a blockchain, enabling digital ownership, trading, and programmable management.

Smart Contract: A self-executing program stored on a blockchain that automatically carries out the terms of an agreement when predefined conditions are met.

SPV (Special Purpose Vehicle): A legal entity created solely to hold a single asset such as a property. Used in tokenized real estate to give investors fractional ownership through tokens.

Stablecoin: A type of cryptocurrency designed to maintain a stable value, typically pegged to a national currency like the U.S. dollar. USDC is a widely used example.

Tokenization: The process of converting the ownership rights of a real-world asset into digital tokens on a blockchain, enabling fractional ownership and instant transferability.

USDC: A stablecoin pegged to the U.S. dollar at a 1:1 ratio. Used in blockchain-based real estate transactions as a stable medium of exchange.

VR (Virtual Reality): Technology that immerses a user in a fully digital environment, typically experienced through a headset.

Web3: The next iteration of the internet, built on blockchain technology, characterized by decentralized applications, digital ownership, and user-controlled data.

www.ingramcontent.com/pod-product-compliance
Lightning Source LLC
Chambersburg PA
CBHW041300210926
51847CB00009B/95
* 9 7 8 1 9 6 2 2 3 2 2 6 5 *